OLYSLAGER AUTO LIBRARY

Earthmoving Vehicles

compiled by the OLYSLAGER ORGANISATION

research by Denis N. Miller

edited by Bart H. Vanderveen

FREDERICK WARNE & Co Ltd
London and New York

Library of Congress Catalog Card Number 72-186749

ISBN 0 7232 1467 0

Photoset and printed in Great Britain by BAS Printers Ltd, Wallop,
Hampshire

INTRODUCTION

This volume concerns a subject known to us all but of which little, so far, has been written. All of us have, at one time or another, stopped to view earthmoving equipment in action, perhaps taking these machines for granted rather than wondering how they were developed or how they work.

The basis of earthmoving is, of course, to move the excavated material from, let us say, point 'A' to point 'B', at the lowest possible cost and, today, in the fastest possible time. Until the introduction of the steam engine, sheer muscle power was the most economical method known to man. Now, the days of the labourer are fast disappearing amidst a welter of mechanical appliances and automation, with the accent upon greater power, higher payloads and improved efficiency.

The modern Plant Manager, unlike his predecessors, must have the most modern equipment at his disposal. If he does not, he must absorb his increased maintenance costs for outdated machinery in his contract tenders, pricing him out of the market and enabling the better operator to step in. Thus, the design and construction of earthmoving machinery, unlike that of certain other heavy industries, is undergoing continual development in order to keep pace with the ever-increasing demands of the modern plant operator and his staff.

Such a vast subject cannot, of course, be dealt with in minute detail in a book of this nature. Instead, we have assembled perhaps one of the most comprehensive selections of earthmover photographs ever published, ranging from the earliest steam-powered machines to the latest push-button controlled diesel-electric giants.

Piet Olyslager MSIA MSAE KIVI

3A : 1901. Manpower was the essential ingredient before mechanization.
3B : 1937. Boydell's Muir-Hill $2\frac{1}{2}$-cu yd 'F'-type.
3C : 1972. The WABCO 200-B 'HAULPAK'* truck, a 1325-bhp giant with a 200-ton capacity ! A 1600-bhp version is also available.

3A   *'HAULPAK' is a registered trademark   3B   3C

ROPE-CONTROLLED BULL- AND ANGLEDOZERS

In November 1904 the historic test of the world's first practical crawler tractor, the converted Holt steam tractor No. 77, took place near Stockton, California. It was intended to solve only a local problem—to replace the horse and conventional wheeled machine in the soft-soiled San Joaquin Valley area by a specially designed appliance capable of operating successfully over such awkward terrain—but it was later to receive worldwide acclaim and open new avenues for the progressive agriculturalists to explore.

In 1906 Holt built an experimental gasoline-powered crawler, and by 1908 the first 100 petrol-driven crawlers were in operation on the Los Angeles Aquaduct project, in the foothills of the Tehachapi Mountains. It wasn't long before these crawlers, and British equivalents, were entering service in all fields. Daniel Best, another pioneer of American steam traction, sold out to Holt in 1908, but his son, C. L. Best, continued to build wheeled tractors, turning to the crawler type in 1913. In 1925 Holt and the younger Best merged to form the Caterpillar Tractor Co.

5B

5A

5C

4A : The first crawler machine was based on Benjamin Holt's steam tractor No. 77. The original rear wheels were replaced by a set of tracks consisting of a series of 3-in × 4-in wooden slats mounted on a linked steel chain. It was figured that these tracks had as much bearing surface as wheels 75 ft in diameter ! Steering was by a special 'tiller' wheel which was later replaced by the 'steering clutch' system whereby one track would be braked while the other continued forward, thus turning the machine.

5A : The first practical bulldozing attachment to be manufactured in any quantity was the manually-operated design built by LaPlant-Choate, an American Company later absorbed by Allis-Chalmers, in 1923. The major disadvantage of this model, here seen fitted to a Caterpillar 'Twenty'

tractor, was the fact that the operator had to leave his cab to raise or lower the blade by hand.

5B : A similar system, with chain adjustment, could be fitted to the earlier petrol-driven Holt crawlers. A 10-ton Holt, seen near Soldier Field, Chicago, in 1924, was of this type, but with blade lifted by a 'second man' who had to dismount on each occasion.

5C : The first bulldozer to be equipped with Robert G. LeTourneau's revolutionary power control system, developed in 1928, was an Allis-Chalmers later that same year. The PCU system, as it was generally known, took power from the engine crankshaft to operate towed equipment or for raising and lowering the blade.

ROPE-CONTROLLED BULL- AND ANGLEDOZERS

6A : Caterpillar's D7 crawler, like its big sister, the D8, was extremely popular immediately after World War II, probably resulting from its excellent war record. This example, equipped with a Cat 7A 'dozer blade, is shown hard at work on a Peoria, Illinois, road-building scheme in March 1946. Specification included a No. 24 front cable control system (the power winch can be seen lower front of the machine) and a 92 hp engine, the latter driving through a 5 forward and 4 reverse speed transmission.

7A : The Cletrac was built by the Cleveland Tractor Company, of Cleveland, Ohio, the British concessionaires being Blaw-Knox Limited. Blaw-Knox also marketed a variety of bull- and angledozer attachments for the Cletrac machine, the example shown being a particularly early, and unusual, rope-controlled model. Power was supplied by a 107 hp diesel engine, providing speeds of between 1.61 and 5 mph.

7B : One of the most popular of all British-produced 'dozers was un-doubtedly the Fowler 'Challenger III' with rope control via an overhead gantry system. Early deliveries featured a Marshall Fowler engine, but because of design problems the bulk of orders had 105 hp Meadows 6DC630 or, later, Leyland industrial power units.

7C : The French ADN130 (c. 1950) was one of the largest crawler tractors of its day, powered by a huge 18-litre engine manufactured under licence from MAN of Germany, requiring a 4-cylinder petrol engine to start it. The ADN blade control system utilized a power winch mounted ahead of the engine with controls placed close to the driver, remote from the winch mechanism.

7B

American and British concerns were leaders in 'dozer design right from the start featuring, initially, manually-controlled blades and, later, electric control, with 'tiller' steering especially popular. During the 1920's, however, Robert G. LeTourneau, an American specializing in the manufacture of earthmoving implements for Holt, Best and, later, Caterpillar prime-movers, developed a new power control system, the prototype of which featured a Hudson car differential, thus widening the the 'dozer's scope still further.

All three control systems featured winch and cable actuation, a method which remained in the 'front line' right up to 1959. One of the last British machines of this type was the revolutionary Vickers 'Vigor', for which hydraulic blade elevation was also offered. The 'Vigor' was developed from the earlier Vickers 'VR'-Series crawlers, first announced in 1951, production of which commenced the following year. The new 'Vigor', however, was just a little too far ahead of its time and production ceased in 1959, only two years after it had commenced.

The introduction of hydraulic 'dozer control in 1938 sounded the slow death knell for the earlier design, although the rope control system can be supplied even now by certain manufacturers.

7A

7C

TRACKED HYDRAULIC 'DOZERS

8A: The 92 hp Caterpillar D7 tractor saw service in all theatres during World War II. This example, with rear-mounted Hyster D7N winch, featuring a 34,500-lb maximum line pull, carried a Birtley blade, and was used by the Canadian Army.

8B: Not only were armoured 'dozers amongst the first machines ashore in the June 1944 landings in France, but they were also employed extensively throughout the Pacific Islands. This heavily armoured model was waiting to move in during the Saipan Operation, also in 1944.

8C: The prototype Blaw-Knox hydraulic 'dozer, with angling blade, made its début during the war. It was based on a Cletrac crawler, suitably modified with front-mounted hydraulic pump driven from the engine crankshaft. Production models, however, featured a 'built-in' hydraulics system.

8D: Production of the Komatsu D50 series commenced in 1947 with the D50A-1, a machine of conventional layout powered by the Company's 60 hp 4D120 diesel engine. On 18th June 1970 the 50,000th D50 rolled off the Company's main production line in Ishikawa Prefecture, Northern Japan.

8A

8C

Hydraulic blade control appeared towards the end of the 1930's. The system originated in the United States, following many years of experimentation and development, and was widely employed during WWII and the Korean War.

In military service the bulldozer was used in both 'soft skin' and armoured forms, and it was under these severe conditions of service that engineers first took real notice of the machine's capabilities. Thus, after the war, demand outstripped supply for about fifteen years, culminating in the development and general acceptance of new high-speed rubber-tyred equipment made necessary by the increasingly large heavy earthmoving contracts essential to modern industrial growth.

8B

8D

TRACKED HYDRAULIC 'DOZERS

9A, 9B : Many of the lighter Blaw-Knox models designed and built during the Fifties were based on David Brown or Fowler crawlers. The former, 9A, powered by DB's own 30—35 hp oil engine, made its début as an independent crawler tractor known as the 'Trackmaster 30' towards the end of 1950, with the B-K 'dozer version appearing the following year. As with many of these early types, hydraulics systems were poorly protected. The Fowler 'VF' Series, 9B, was similar, mechanically, to the famous 'Field Marshall' tractor. This was the B-K angledozer version, featuring a 9-ft. blade with 3-piece reversible cutting edge.

9C : The hydraulic version of the Vickers 'Vigor' featured a 'back-to-front' hydraulics system, so designed to eliminate much of the wear and tear encountered by this type of machine. An exclusive feature of the 'Vigor' was the supercharged Rolls-Royce diesel of 180 hp driving through a power transmission or torque converter. Another unique feature was the contour-clinging track layout incorporating self-articulating suspension.

9A

9B

9C

10A : One of the most powerful crawler machines currently offered by International Harvester is the Series 'C' TD25, a 285 hp turbocharged machine powered by an International DT-817B engine and incorporating a unique power steering system. This eliminates the familiar 'physical jerks' caused by alternated braking and power application to the tracks in order to accomplish a turn.

10B : In recent years the US Army has developed a system whereby smaller machines, such as this Series 'C' Caterpillar D4 tractor, can be dropped behind enemy lines to speed airstrip construction or road clearance. Using six large parachutes and a special pallet loaded with cushioning material, such a method has already proved relatively successful in the Vietnam conflict.

10C : Under extreme battle conditions even the armoured tank has been fitted out with 'dozer equipment. This US Army 90-mm gun tank type M47, for example, carried a heavy blade both for earthmoving and for clearing obstacles from its path.

10D : Some of the larger projects may require a special tandem 'dozer, consisting of two identical crawlers coupled one behind the other, with a single blade in front. Sometimes, however, even more exotic schemes have been devised. The Caterpillar SXS D9G tractor, for instance, is designed for extra wide single cuts of up to 24 ft. It consists of two D9G machines coupled side by side with a single blade raised and lowered by four hydraulic rams. The total power output is approximately 770 hp, controlled from the tractor on the right of the picture.

10A

10C

10B

10D

TRACKED HYDRAULIC 'DOZERS

11A : A machine of unconventional appearance is the Russian Belaz D-572, of which there are many examples outside 'Iron Curtain' countries. The weatherproof cabin (essential for use in many parts of the Soviet Union) is mounted centrally, with power supplied by a 4-stroke V-12 of 350 hp. The blade width of the standard machine, which is based on the DZT-250 industrial tractor, is 16 ft.

11B : A popular attachment for this class of machine is the rear-mounted ripper. This 3-prong root ripper is mounted on a 140 hp Hanomag K12X machine manufactured in West Germany by Rheinstahl Hanomag AG. The engine compartment is totally enclosed and dust-sealed.

11C : Believed to be the world's largest and most powerful crawler 'dozer, the Allis-Chalmers HD-41 is powered by the largest diesel in any crawler, a Cummins VT-1710-C turbocharged V-12 diesel developing some 524 hp. It is capable of shifting 24 cu yd of material in one pass on level ground and has a drawbar pull of 185,000 lb. The first prototype appeared in 1965, powered by an experimental 540 hp V-8. By 1966 the Cummins V-12 had been selected, and further prototypes constructed and tested before production stage.

11A

11B

11C

WHEELED 'DOZERS

12A: LeTourneau's earliest rubber-tyred bulldozer was capable of travelling at up to 15 mph in either direction, having the same four speeds forward and reverse. The steering principle was identical to that of a crawler machine and the use of four speeds in either direction led to faster returns for pushloading and 'dozing.

12B: Bomford & Evershed's first 'dozer attachment was announced in 1954. It was fitted to the popular Ferguson agricultural tractor and later produced under licence by Harry Ferguson (now Massey-Ferguson). The tractor featured a patent overload skid release system which would cut out the power if the machine lost traction during the 'dozing operation.

12C: Amongst the approved for the Land-Rover is J. B. Howie's hydraulic 'dozer blade manufactured by Atkinsons of Clitheroe Limited.

12D: The David Brown 'Highway Tractor' is aimed at local authorities and industrial concerns rather than agricultural users. The standard front-end loader bucket can be replaced by a small angledozer blade by means of three quick-release pins.

12B

12C

12A

12D

WHEELED 'DOZERS

13A

13B

Bull- and angledozer attachments for agricultural tractors made their appearance about 1930. Initially, they were developed for agricultural use but, inevitably, their capabilities were quickly noticed by the building and construction industries. Two major advantages over the crawler 'dozer were that the wheeled type could operate at higher speeds and it could travel considerable distances on the public highway under its own power.

One of the first true rubber-tyred 'dozers was designed and built by LeTourneau in 1946. This was the first machine ever to be designed specifically for use as a bulldozer, opening yet another market to the heavy plant manufacturer. From 1947 until 1953, 453 of the Model 'C' 'Tournadozer' were produced, with the Super 'C' introduced around 1952. Hundreds of these LeTourneaus saw active service in the Korean War and proved themselves beyond any shadow of doubt. The high-speed earthmoving field was here to stay.

Since then many other manufacturers have introduced similar machines, often following closely the lines of the wheeled loading shovel. One of the most famous of these is the International Hough 'Paydozer' series, announced by the Frank G. Hough Company, of Illinois, soon after their take-over by International Harvester in 1952.

13C

13A : Many of the leading earthmoving equipment manufacturers offer special compactors equipped with 'dozer blades as standard equipment. These machines, with their unusual steel wheels, have a similar effect upon soft materials such as earth or sand as a diesel roller has upon a newly laid tar-macadam road surface. This is the Caterpillar 815, a 170 hp model with single-lever powershift planetary-type transmission and an articulated frame. It is the smallest of three in the Caterpillar range, the others being the 300 hp 825B and the 400 hp 835.

13B : Muir-Hill's industrial tractor range, consisting of the 101, 110 and 121 models, is available with a specially designed twin-ram power 'dozer

attachment with angling blade, designed, manufactured and marketed by Bomford & Evershed Limited. This, the 4-wheel drive 110 series, like the others in the range, has been developed from components designed specifically for the Company's range of wheeled loading shovels.

13C : The International Hough D-500 'Paydozer' is claimed to be the largest wheeled bulldozer in the world. It is particularly popular as a 'pusher' unit for the largest motor scrapers, and is itself fitted with a special 'push' plate, at the rear, for double-banking on a gradient or in soft conditions.

Crawler loaders were introduced about 1930, using engines of some 15–20 hp capacity. These early cable-operated types were a development of the bulldozer, utilizing a crude bucket of approximately ½-cu yd volume mounted in place of the 'dozer blade.

A pioneer in this field was the Trackson Company, of Milwaukee, Wisconsin, designers and manufacturers of a number of novel types based on Caterpillar crawler tractors. Known as 'Traxcavator' loaders, these machines became so popular that the Caterpillar Tractor Co. acquired Trackson in 1951, applying the trademark 'Traxcavator' to all following crawler loaders.

The earliest hydraulic types did not appear in any quantity until after World War II but, like the bulldozer, earlier versions underwent extensive operational trials during the war. Most famous of these was the International TD9, better known in later years as the International Drott. It was this machine which played a large part in revolutionizing the contracting plant hire industry.

15A

15B

15C

14A : The Anderson Brothers, of Port Richmond, Staten Island, developed an interesting 'shovel loader' early in 1932. It was based on a Caterpillar 'Fifteen' crawler tractor and consisted of a ½-cu yd bucket which was loaded from the rear i.e. the machine was reversed into the material to be moved) and transferred over the tractor into the truck. Unlike other designs, the loader was not required to manoeuvre round, sometimes through 180°, to deposit its load.

15A : It was claimed that the Trackson 'Traxcavator', as based on the Caterpillar D4 machine, could work an 8-hr day on less than 2 gallons of fuel. The operation was based on a similar principle to the fork-lift truck, whereby the bucket was raised up a vertical frame and the machine manoeuvred round to tip the load.

15B : Wartime versions of the International TD9 were frequently referred to in military circles as 'sand scoopers'. This early model saw service on Guadalcanal in February 1943 and was the basis of many future International Drott crawler loaders.

15C : The Case 450 with Drott '4-in-1' bucket appeared on the British market in 1967 and was soon acclaimed the fastest 1-cu yd crawler loader in the country. A revolutionary feature at that time, now common to a number of machines, was the torque converter transmission, known in this instance as the Terramatic system, which some plant men regarded as way ahead of its time.

WHEELED LOADERS

The wheeled loader, like its counterpart in the 'dozer field, was a development of the agricultural tractor. It did not become a practical machine until the introduction of LeTourneau's power control system in 1928. Even so, there were a few hand-controlled types before this.

Rope-controlled models continued to be popular up to the late Thirties, but concerns such as the American Frank G. Hough Company (now the Hough Division of International Harvester) were quick to realize the potential of the hydraulics system and soon had a small prototype fleet in military service.

16A

16C

16B

16D

16A : E. Boydell & Company Limited, of Manchester, introduced their Muir-Hill loader range in 1929. The first model, a ½-cu yd rope-controlled unit, was based on a modified version of the 28 hp Fordson tractor.

16B : One of the most popular of early British types was the Chaseside 'Hi Lift Major' rope-controlled model. Again this was based on the Fordson agricultural tractor, featuring a similar loading arrangement to the earlier Muir-Hill design but with a more advanced control mechanism, rear-winching system and counterbalance weights.

16C : The Fordson was also the basis of the JCB 'Majorloader', developed in 1949 by J. C. Bamford Limited. Unlike the Chaseside, the 'Majorloader' was of fairly simple construction, affording greater operator visibility over many competitive machines of the day.

16D : Cie des Automobiles Industrielles Latil, of Suresnes, Seine, France, built many varied and often unusual commercial and industrial vehicles until production ceased in 1955. One of their later models was this unique hydraulic front-end loader with four-wheel steering.

17A : A small capacity industrial loader, known as the LH-1, was introduced by E. Boydell & Company (Muir-Hill) in 1952. With a mere 10½-cu ft bucket capacity, one of its most unusual features was the rear engine with centrally-mounted operator's position.

17B : The Weatherill Hydraulic 4HTW two-way 'Overloader' was another interesting design. This was based on later versions of the Fordson tractor, utilizing three pairs of rams to facilitate loading from the front and dumping to the rear. It is claimed to have been the first hydraulic overloader to be introduced in the United Kingdom.

17C : The first JCB 'Loadall' front-end loader appeared in 1956, a worthy successor to the Mk. I JCB of 1954. The basis was, once again, a Fordson. For the first time a simple system of fitting an hydraulic loading shovel to a standard agricultural appliance had been perfected.

After the war, British manufacturers challenged America's lead in wheeled loader development. The Chaseside Engineering Company Limited and W. E. Bray & Company Limited, for example, had been manufacturing loading equipment, principally for the Fordson tractor range, for a number of years. They were now joined by numerous other concerns all striving towards a common end—to develop a successful range of hydraulic wheeled loaders built exclusively for that purpose.

The Bray 'Hydraloader', a ⅞-cu yd appliance based on the Fordson 'Major' tractor, with which it was sold as a complete outfit, was claimed to be the first all-British hydraulic wheeled loader when announced in 1949. This was soon joined by the Weatherill 2H of 1950, of which more than a thousand were sold. The first all-British four-wheel drive machine was the 'Brayloader' BL430 which, like the 'Hydraloader', was built at the Feltham, Middlesex, works of W. E. Bray & Company.

17B

17A

17C

18A

18B

18C

18A : The first new model to leave Weatherill's new Welwyn plant was the 12H in 1958. This was the Company's most successful two-wheel drive machine, with total orders in the region of 3000. Its unusual contours, with centrally placed cabin, belie the fact that even this was essentially a Fordson tractor.

18B : A $\frac{7}{8}$-cu yd four-wheel drive loader of unusual design was announced by Muir-Hill in 1958. This was the 4WL which, with Fordson engine and counterweights slung low between the wheels, was claimed to have an extremely low centre of gravity.

18C : Assets and patents of the Merton Engineering Company were acquired by Whitlock Bros in 1968, production of the full Merton range being transferred almost immediately to Whitlock's Great Yeldham factory. By this time the rear-engined loader was almost universal, with the engine counterbalancing the laden weight of the bucket. In common with other Whitlock 'Merton' loaders, the 120B 4 x 4 model was identical to its all-Merton predecessor.

18D : The manoeuvrability of the articulated loader is superbly illustrated by the German Zettelmeyer Model KL 30. Power is supplied by a Deutz V-8 4-stroke diesel of 240 hp. This is transmitted via a single-stage torque converter and 4-speed full-reversing power-shift transmission to the four driving wheels.

18D

WHEELED LOADERS

Since the mid 1950's, the wheeled loader market has been led by American or American-owned manufacturers, with British types a close second. Apart from the latest Muir-Hill range, the majority of wheeled loaders currently available are of very similar layout, often with four-wheel drive, some with articulated chassis and still more with four-wheel steering, the latter having been introduced by the Clark Equipment Company (Michigan) in 1965.

The latest development to be tried in this field is the Caterpillar 'Dystred', a compromise between the rubber tyre and the crawler track. Whilst wheeled earthmoving plant speeded up materials handling operations, tyre life expectancy was cut severely by the considerable wear and tear incurred by such arduous operations. The 'Dystred' system is aimed at combating this and, at the same time, provides improved traction. Another Caterpillar venture is the '988 Carry Loader', with the unusual attributes of buckets both fore and aft. The idea behind this is to increase the efficiency of the load and carry cycle.

19A

19B

19A : Eimco TMD, a Division of Envirotech, specializes in the design and construction of earthmoving and materials handling machinery, principally for use in underground mining. Typical of these revolutionary designs is the low-height 915 LHD front-end loader. This is only 65 inches high and yet it delivers a 5-cu yd bucket capacity. To obtain such a low height, the operator is positioned to one side of the machine with the Deutz F8L 714 diesel engine located low down at the rear.

19B : Current Muir-Hill loaders are of an entirely different layout to other makes. This, the military version of the A5000, illustrates the major differences, such as the long, single, centre-mounted lifting beam, improving balance, overall reach and tipping height ; offset safety cab providing maximum visibility ; and, on this occasion, an Owen/Drott '4-in-1' loader bucket manufactured exclusively in the British Isles by Rubery Owen & Company Limited, Darlaston, Staffs. This bucket is so called because of its ability to load, level, pile or scrape.

19C : When WABCO builds a big machine they do mean big ! This is the recently introduced Model 1200 loader, a fully articulated four-wheel drive machine of 30,000-lb capacity, which is rapidly replacing the excavator for certain applications.

19C

MULTI-PURPOSE LOADER/BACKHOE MACHINES

By 1952 wheeled and crawler loaders were rapidly becoming the mainstay of the light- and medium-duty earthmoving fields. It was not surprising, therefore, that further uses were found for these machines. One such idea, which later revolutionized the building trade, was the backhoe or excavator attachment, designed for rear mounting in place of the usual counterweights.

The backhoe is an hydraulic attachment, normally actuated by a second set of controls located behind the operator. Frequently, the operator has

two seats, one for driving and when using the front-end loader, and the other, located higher to provide better control and visibility, for working the backhoe.

In Britain, early designs were developed by Whitlock Bros in 1952 and by J. C. Bamford Limited two years later. Other manufacturers, basing their designs on similar principles, announced their own models soon after.

20A : Whitlock's 'Dinkum Digger' was developed in 1952, fast becoming the leading multi-purpose earthmoving machine in the industry. Based on a Fordson tractor, and of particularly robust design, it was, in fact, the first tractor-mounted hydraulic excavator to be seen in the United Kingdom. The machine shown here was not the very first 'Dinkum Digger', but was an early example of this range, being produced in 1953.

21A : As early as 1951 J. C. Bamford Limited was mounting a specially designed backhoe unit on its front-end loader conversion of the Fordson 'Major' tractor. The hydraulic pump was driven off the engine crankshaft with fluid stored in the rectangular tank on the right-hand side.

21B : The J. I. Case Company Limited offers a variety of backhoe equipment for mounting on any of the Company's crawler range of front-end loaders. Here, a Case 750 crawler loader, with Drott '4-in-1' loader bucket has been equipped with one of these backhoe units. The 750 is powered by a 70 hp engine and has a capacity of $1\frac{1}{4}$ cu yd.

21C : The Swedish BM-Volvo GM410 was recently introduced into the United Kingdom to compete against existing types. Of particular interest to British operators is the spacious weatherproof cabin typical of a machine originating so close to the Arctic Circle.

21A

21B

21C

MULTI-PURPOSE LOADER/BACKHOE MACHINES

22A : The Whitlock 370 features the novel 'Swinglock' method of backhoe offsetting. This entails slewing the kingpost to the required position using hydraulic rams and then locking it in place. The 1-cu yd front-end bucket can be of the conventional type or of the patent '4-in-1' design.

23A : Popular behind the 'Iron Curtain' is the Russian Z-153 design. This is not strictly a backhoe unit as the rear-mounted excavator works on a similar principle to the rope-operated 'crowd shovel'. This action is obtained by strategic placing of two hydraulic rams.

23B : The British Ford '13-Six' offset digger/loader, introduced in 1970 by Ford Tractor Operations, is a very popular machine. It is claimed that this could be due to a revolutionary new idea known as 'Auto-Dig', featuring a special sensing device which can be switched in to control the digging cycle.

23C : The Hy-Dynamic Company's Model 200 is claimed to be the largest rubber-tyred backhoe/loader in the world. A 163 hp Detroit diesel provides power for all movements, steering is through the rear wheels only, and four-wheel drive is another feature. The maximum digging depth of the backhoe is 20 ft and loader bucket capacity as standard $2\frac{1}{4}$ cu yd.

23B

23A

23C

ROPE-OPERATED EXCAVATORS AND DRAGLINES

24A : The Otis American Steam Excavator, announced in 1838, was one of the first mechanical shovels used on early American railroad projects. It was mounted on a wide-gauge track and equipped with a chain hoist mechanism and single-cylinder engine.

24B : Established in 1820, Thomas Smith & Sons (Rodley) Limited, of Leeds, England, introduced their first powered excavator in 1887. This was basically one of the Company's already popular 3-ton rail-borne steam cranes fitted with a crude shovel attachment.

24C : The Marion Model 75 steam shovel was in production from 1901 until 1912. With a 4-cu yd capacity, this design was built in such a way as to be readily transportable on a standard gauge railway track, simply by lowering the jib and swinging the stabilizers in close to the superstructure. It was, therefore, especially popular for railroad construction work. It is interesting to note that the superstructure was rigidly mounted on the truck base, while the shovel itself was located on its own limited-slew turntable. The bucket door was opened and closed manually.

24D : The first full-slew long-range stripping shovel was also a Marion. This was the 3½-cu yd Model 250, produced from 1911 until 1913. Designed for use on opencast mining sites, this was also a rail-borne machine, using, on this occasion, a pair of parallel tracks.

The great railway 'boom' of the last century saw the introduction of the first true mechanical excavators. Messrs Eastwick & Harrison, of Philadelphia, were pioneers in this field, introducing their limited-slew type Otis American Steam Excavator which was popular for railroad development as early as 1838. A few years later (*c.* 1874), on the other side of the Atlantic, Ruston Proctor & Company introduced their famous 'Steam Navvy'. It was assembled to the patents of James Dunbar, and it, too, found employment on various railway projects, although one of its first major contracts was the Manchester Ship Canal in 1887.

24C

24A

24B

24D

ROPE-OPERATED EXCAVATORS AND DRAGLINES

Further developments came thick and fast. New machines were introduced by the Vulcan Steam Foundry, Bucyrus of Ohio, the Osgood Company and a triumvirate comprising Henry M. Barnhart, a supervisor on the Chicago & Atlantic Railroad, George W. King, a manufacturer of agricultural equipment, and Edward Huber, an inventor whose name lives on in the Huber range of construction equipment. Together they formed the Marion Steam Shovel Company (later renamed the Marion Power Shovel Company) in 1884, which was destined to become one of the world's leading manufacturers of abnormal size earthmovers.

Without exception, these early types were of the limited-slew design,

the first recorded full-slew machine being assembled by Whitaker of Leeds in 1884. Until 1922 the majority of these machines were rail-mounted, a few had steel road wheels and one or two featured the new (for mechanical excavators) crawler track system.

The giant excavator or dragline, because of its proportions, is largely restricted to the opencast mining field where excavations are more or less permanent. In many instances these machines are assembled on site by field engineers sent out by the manufacturers, a task which, in the case of the largest designs, can take many months.

25A

25C

25B

25A : The Speeder Machinery Company, of Leon, Iowa (purchased by the Link-Belt Company in 1939 to form Link-Belt Speeder), was founded in 1919, their first product being 'a mechanical loader for handling roadside gravel and for general excavating'. It was powered by a small petrol engine and featured chain-driven front wheels and a 'scoop' or 'drag' excavating attachment.

25B : Yet another interesting design was that produced by the Northwest Company in the USA during the Twenties. Seen here on a new highway site in 1925, this model of the period bore a strong resemblance to more modern machines of this type. The petrol-engined Best crawler and 'double-bottom' dump outfit was typical of heavy earth-hauling equipment of the day.

25C : The American Hoist & Derrick Company, founded by Oliver Crosby and Frank Johnson in 1882 as the American Manufacturing Company, based in St Paul, Minnesota, specialized in rail-borne steam cranes for a number of years before adding a steam excavator to its range. This was succeeded during the early Thirties by an internal combustion-engined type, also with face or 'crowd shovel' equipment.

ROPE-OPERATED EXCAVATORS AND DRAGLINES

26A : The British Priestman excavator remains a best-seller in the United Kingdom and abroad. Two of the Company's most famous models were the 'Panther' (in foreground) and the 'Cub', seen here early in 1940. These were available with various types of excavating or lifting equipment but, as shown, carried Priestman's own 'Navvy Shovel' attachments.

27A : Largest in the current range available from Thomas Smith & Sons (Rodley) Limited is the Rolls-Royce powered E.4000 2½-cu yd dragline. Draglines work on a similar principle to the motorized scraper in that they strip the top layer of soil by dragging the bucket, swinging round to deposit the material elsewhere. Face shovel or backhoe attachments are also available.

27B : Europe's largest 'walking' dragline, owned by Derek Crouch (Contractors) Limited, can be seen at the National Coal Board Opencast Executive's 2000-acre site at Radar North, near Widdrington, Morpeth, Northumberland. The machine, a 3000-ton 1550W Bucyrus Erie, cost £2.3m and is capable of moving 5500 tons of material per hour, coping with 65 cu yd (100 tons) per bite !

27C : Marion's Model 6360 giant stripping shovel must surely be one of the largest land machines in the world. It is equipped with a 215-ft boom and 180-cu yd bucket. The shipping weight is approximately 10,825 tons and the huge drum at the front of the machine carries the cable which supplies electric power for all movements.

27A

27B

27C

BUCKET WHEEL EXCAVATORS

The bucket wheel principle is best understood by studying the accompanying illustrations. Certainly, the great attraction of this system is that the wheel is excavating continuously, depositing the excavated materials via a system of conveyors onto an adjacent dump or into waiting trucks, thereby cutting down on the amount of 'dead' time in the excavating cycle. This continuous principle also results in lower costs and less wear and tear than for the conventional excavator.

Broadly speaking, the bucket wheel was developed in Europe with the most popular designs emanating from Germany, notably by Krupp and Lübecker Maschinenbau. In more recent years, Anderson Mavor Limited (formerly trading as Mavor & Coulson Limited), of Glasgow, have developed their E10 model for use in the general excavating field. This is a complete departure for this class of machine as, previously, it has been employed almost exclusively on opencast mining and quarry work.

possibilities of the bucket wheel system as applied to smaller machines. A pilot model was mounted on an old Coles crane chassis in 1965 with the first prototype designed and built by M. & C. appearing in 1966. This machine, of 4 × 4 configuration, is shown undergoing evaluation trials prior to building the third, and much improved version.

29C : This Mavor E10, with a theoretical output of some 470 cu yd per hour, was the second production machine and the eighth built by Anderson Mavor Limited (trading as Mavor & Coulson Limited until April 1971). So far, it has been used extensively on a reservoir project in Buckinghamshire, loading excavated material into Cat 631B wheel-tractor scrapers. Power comes from a Leyland UE.680 diesel developing 170 hp, whilst options include Rolls-Royce and Cummins units.

29A

28A : One of the world's largest bucket wheel excavators was assembled on site by Lübecker Maschinenbau, an associate of Orenstein & Koppel. This vast machine, known as the B.1166, travels on huge crawler tracks and has a single main boom with counterbalance and conveyor booms projected out to the rear. The bucket wheel operator's cabin is located at the head of the main boom to provide a good view of the excavation cycle, with a second operator controlling the conveyor boom and tipping cycle. The immense size of this machine can be gauged by the bulldozer to the left of the bucket wheel.

29A : This German Krupp bucket wheel works on a similar principle but is on a much smaller scale. The National Coal Board Opencast Executive's example at their Radar North site excavates a face up to 40 ft high, elevating the material to ground level and thence by dump conveyor to the tip area.

29B : In 1963, Mavor & Coulson Limited began looking into the

29C

WHEELED AND TRACKED HYDRAULIC EXCAVATORS

30A

30B

30C

It is interesting to note that the hydraulic excavator was thought of even before the first 'Steam Navvies' appeared. One pioneer in this field was Sir W. G. Armstrong who developed a number of variations during the 1880's.

The hydraulic excavator 'boom', however, did not occur until after World War II. As engineers were developing the hydraulic bulldozer and loader, so others were experimenting with the application of hydraulics to actual excavating equipment. Leading the field was

30D

30A : Poclain's TU Model was one of the earliest types of hydraulic excavator in Europe. The prototype was a 2-wheeled trailer model drawn by a tractor, and driven off the tractor pto. It was developed in 1951.

30B : In 1956 Brödr. Söyland A/S, Bryne, Norway, designers and manufacturers of specialist engineering plant, unveiled a totally new concept in hydraulic excavation. Known as the 'Broyt', it 'married' the conventional with the unconventional by featuring no-wheel drive. Instead, it propelled itself forwards, backwards, sideways, or even lifted itself clear of obstacles by means of its powerful hydraulic bucket arm, available with shovel or backhoe attachment. Early production models featured 3-cylinder Ruston or Bolinder engines or a 4-cylinder Perkins unit.

30C : An early, but unfortunately unsuccessful, entry into the hydraulic excavator field was Blaw-Knox Limited, of Rochester, Kent, with the 'Hydrascoop' of 1959. This featured a complicated system of rams and hydraulic lines, and although of greater capacity than many other types, was severely handicapped by its weight.

30D : The Koehring 505 'Skooper' was similar to the 'Hydrascoop' and equally unsuccessful. It was built by Messrs NCK-Rapier Limited who now concentrate upon crane, rope-controlled excavator and dragline production.

WHEELED AND TRACKED HYDRAULIC EXCAVATORS

Frenchman Georges Bataille who, using American techniques developed during the war, introduced his first all-hydraulic design, the TU, in 1950, naming it after the nearby village of Poclain. This is believed to have been the first all-hydraulic excavator built in Europe. It was trailer-mounted and shown for the first time at the annual Agricultural Machine Exhibition, Paris, in October 1951.

Since 1955, interest in this type of machine has increased more than tenfold, and experiments are now proceeding into the even higher weight classes—where it has been said that hydraulics could never equal rope-controlled methods.

31A

31B

31A : The Link-Belt Speeder Division of the American FMC Corporation manufactures a vast range of crane and excavator equipment. In 1936 the Link-Belt Machinery Company which, three years later, purchased the Speeder Machinery Company outright, introduced power-hydraulic controls to the excavator industry. The first all-hydraulic excavator, however, did not appear until the mid 1960's, before the Company merged with the FMC Corporation. This was the LS-5000.

31B : Pioneers of hydraulic excavation in the British Isles were the Hy-Mac Company, formed by Peter Hamilton Equipment Limited to design, build and sell a complete range of rubber-tyred and tracked full hydraulic excavators. The first model, the ½-cu yd HM-480, was introduced in May 1962, eighty machines being completed in the first year. Total production for all types is now in excess of 180 per month. Two similar machines, a wheeled HM-610 and a tracked HM-580, are seen here with face shovel attachments.

31C : The German Hanomag range is very similar to the Hy-Mac. This M60 four-wheel drive model also has hydraulic steering.

31C

32A

32A : The Swedish Akerman range was introduced between 1965 and 1966, appearing in Britain some two years later. The largest is the H25, powered by a 276 hp Scania DS11 R40 engine.
32B : Instead of designing and building their own range of hydraulic machines, Whitlock Bros, of Great Yeldham, Essex, who had already made their mark with a range of dual-purpose tractor loader/backhoe units, signed an agreement with Johs Fuchs Kg., of Ditzingen, West Germany, whereby Whitlock could manufacture and market in the United Kingdom, and elsewhere, selected equipment from the Fuchs range. One of these was the originally Deutz-powered Whitlock-Fuchs 50R, a full-slew hydraulic design of standard layout, now available with optional Ford or Perkins engines.
32C : Another well-known name in earthmoving circles is Yumbo of France, founded in 1956 to market the very first self-propelled 360° continuous rotating hydraulic excavator on the French market. Models such as the Y35, a 30 hp machine shown here, incorporate a unique turntable levelling device which permits levelling of the upper structure up to 11° either side. The device is regulated by a single hydraulic control acting upon two rams. Much of this Company's capital is now tied in with International Harvester.
32D : The American Hoist & Derrick Company has also joined the hydraulic equipment league with a new full-hydraulic design known as the Model 25. Available with a 1- or 1¼-cu yd backhoe bucket, this machine is powered by a Detroit diesel of 185 hp, with a V-8 Cummins as alternative. Full hydrostatic-drive crawler tracks are standard.

32B

32C

32D

WHEELED AND TRACKED HYDRAULIC EXCAVATORS

33A

33A : The truck-mounted Link-Belt Speeder HC-2000 'RotaScope-Hoe' incorporates a full-slew excavator unit with double-extension cylindrical boom, raised and lowered by twin rams, but with rope actuation of boom length. This type is particularly rare.

33B : The Poclain EC1000, a 140-ton giant, is one of the largest full hydraulic excavators in the world. Power comes from three Detroit diesel V-8's, developing some 780 hp in all, and backhoe buckets from 4 to $6\frac{1}{2}$ cu yd are offered. With shovel equipment, capacities vary from $5\frac{1}{4}$ to $10\frac{1}{2}$ cu yd.

33B

C

TRUCK-MOUNTED AND SIMILAR EXCAVATION ATTACHMENTS

34A : The Allen 'Dorset' model featured an independent 20 hp David Brown petrol/paraffin engine for controlling bucket movements. The basis of the machine was a short-wheelbase Bedford 5-tonner, and backhoe or shovel buckets could be fitted.

34B : The Mercedes-Benz 'Unimog' is offered with a variety of excavating appliances, including the Klaus Type 205 rotating backhoe appliance. This is a 360° full-slew machine incorporating four stabilizer jacks, operating off the vehicle's multi-purpose hydraulic system and pto-shafts. The 'Unimog' first appeared in 1948 and has rapidly become a leader in the multi-purpose on/off-highway field.

34C, 34D : Thwaites Engineering can now offer an interchangeable backhoe attachment for their light site dumper chassis. One of the Company's earlier hydraulic excavation aids, however, the 'Ninedeep' digger, 34D, was a self-contained machine incorporating an unusual 'flat-top' base and large weatherproof cab. The 'Tusker' digger/dumper, 34C, is capable of excavating to a depth of 8 ft 6 in with bucket widths from 1 to 2 ft. It is claimed that the excavating attachment can be replaced by a 30-cwt dumper skip in less than three minutes.

34A

34C

34B

34D

TRUCK-MOUNTED AND SIMILAR EXCAVATION ATTACHMENTS

35A : The Series II Wakefield self-loading dumper was a very rare four-wheel drive machine designed and built in the United Kingdom by British Jeffrey-Diamond Limited, using a 100 hp Leyland AU.350 power unit. The basic design was developed for use in confined areas where a separate loading machine could not be provided. Thus, many of these machines were, and still are, used on underground mining activities. Features included 2-way steering and controls of the 'shuttle' type, a 10-speed transmission for use in either direction and three basic body styles. The shovel had a capacity of 1 cu yd with struck body capacities ranging from 4.3 to 5 cu yd.

35B : The self-loading tipper is now a most necessary part of the world's construction industry. One of the first manufacturers to cater for this market was Atlas, a German company, who mounted this grab attachment on a four-wheel drive Henschel. The Company's 6000 loader, which has a capacity of 5 tons at 14-ft radius, has even been applied to Cat D8 crawlers for pipe-laying operations.

35C : A pioneer in the use of this equipment in Britain was F. Webb Limited, a West London tipper operator. The Company now has an extensive fleet of self-loaders, the majority being Commer 'Maxiload' types with Hiab loaders. One of their earlier machines, however, was a Ford Thames 'Trader II' with Hiab 'Speed-Loader' $\frac{5}{8}$-cu yd clamshell attachment.

The advantages of mounting hydraulic excavation attachments on truck chassis were not, at first, fully realized. An early design, known as the 'Dorset' and mounted on a normal-control Bedford chassis, was developed by John Allen & Sons (Oxford) Limited. The idea behind it was to provide a small excavator of considerable mobility for tackling varied and scattered work—a machine, in fact, which would be particularly suitable for local authority work.

This type of vehicle is rare, however, the most common types in this field being light dumper chassis carrying tractor type backhoe units, often interchangeable with dumper skips.

In recent years one vehicle type in particular has become more and more popular. This is the self-loading tipper. Such equipment has in many cases eliminated the need for a separate loading shovel to be in attendance on site when excavated or other surplus materials must be removed.

35A

35B

35C

TRENCHERS

The mechanical trencher introduced a fast, efficient method of excavating continuous trenches, using the dredger principle of the endless bucket chain. First efforts to design such a machine came during the Twenties when the internal combustion engine was fast becoming the 'norm' in construction and civil engineering circles.

The Cleveland Trencher Company of Cleveland, Ohio, the Barber-Greene Company of Aurora, Illinois, and the Buckeye Manufacturing Company of Anderson, Indiana, led the field with their early tracked models and, even now, both Cleveland and Barber-Greene build a comprehensive selection of types for service under all conditions. The British Allen range, another leader in this field and a former UK distributor of the Buckeye series, now consists solely of the 'Landraner' and 'Chanedrane' machines, both designed for agricultural land drainage work and equipped for simultaneous laying of clay tile or plastics piping.

36A   36B

36A, 36B : Early Buckeye trenchers of the 1920's featured crude petrol engines, solid steel wheels and tracks. For instance, this early model seen undergoing development trials was powered by a 4-cylinder petrol unit. Trenching equipment consisted of a continuous bucket wheel, chain-driven from the engine, unloading onto a moving rubber belt which could be raised or lowered by a hand-winching system.

36C : From 1930 until 1970 the mainstay of the Allen series was the 16-60 of which the first model, the Series 25, was in production right up to 1946. Power for the last of the series was provided by a 66 hp Dorman diesel and the machine could cut trenches from 16 in to 5 ft in width and up to 14 ft in depth or, with the aid of a special extension unit, up to 16 ft in depth.

36C

37A : Both the German and French Armed Forces use the 'Matenin-Batignolles' KX609 four-wheel drive trencher for excavating work. It is equipped with a front winch, bucket chain system and one, or even two, discharge conveyors.

37B : The Vermeer trencher is just one of the many small trenchers now popular in the United States. The 60 hp M-460, like other Vermeer models, is fitted with high flotation tyres and a front-mounted 'dozer blade. This unit also features an articulated backbone and 4-speed transmission. The digging chain is the heaviest offered by the Vermeer Manufacturing Company and can excavate to a depth of 72 in.

37C : Prominent in Dutch land reclamation work is this mechanical ditcher, designed by Mr G. T. Rodenhuis, and first in service in 1965. Most important features are its light weight and low-pressure tyres, essential for a machine operating in areas where the pressure exerted at ground level must not exceed even that of human feet ! Power, supplied by a Leyland 0.900 diesel engine, is transmitted either to the digging blades only or to these and all traction wheels, drive to the latter being of the electric type. It is interesting to note that the digging blades can also be replaced by grader or scraper attachments, depending upon the operator's requirements at the time.

In recent years a 'mini trencher' revolution has spread from the United States. These machines, both wheeled and tracked, are aimed notably at the public utilities such as telephone companies, electricity boards, etc., or their contractors, and are designed for shallow depth excavating work. Many are also equipped for laying continuous plastic piping or clay drainage pipes simultaneously, making the purchase of such machines a highly economical proposition.

The use of larger machines has now more or less died out. Too many modern trench excavating contracts require precision work to avoid many of the numerous services hidden beneath our towns. The hydraulic excavator or digger/backhoe machine is far better suited to this type of work.

37B

37A

37C

SITE DUMPERS

These were relative late-comers to the earthmoving scene, the first types not arriving until the late Twenties. They were really intended for general site work as opposed to specific earthmoving jobs, and were a considerable improvement over the wheelbarrow or narrow gauge rail-borne dumper, both previously universal methods.

The real demand for the light site dumper came about during the post-World War II building 'boom' when the necessity for a light, highly manoeuvrable, 'go anywhere' machine became apparent.

E. Boydell & Company Limited (Muir-Hill) introduced their first dumper, a 2-cu yd steel-wheeled machine, in 1927, following this with a series of dumpers gradually increasing in size. Barfords of Belton Limited introduced their 3-wheeled skip dumper in 1946 and this, too, was relatively successful. Other companies, such as Chaseside Engineering, entered the site dumper field on a relatively small scale but eventually pulled out to concentrate on other types of machine which, in Chaseside's case, was the wheeled loader.

These early models featured manual tipping systems but, with the introduction of Thwaites Engineering's first design, another 3-wheeler, with hydraulic tipping control, the pace quickened and hydraulic tipping, four-wheel drive and, now, hydrostatic transmission came to the fore.

39B

38A: The first Muir-Hill dumper was similar in many ways to an agricultural tractor with a central transmission unit acting as the main frame member and extra large diameter steel driving wheels. The all-steel bucket, a 2-cu yd gravity tip model, was of the scow-end type and this, plus the rear-facing operator's position, were to remain features of the site dumper for many years.

39A: In 1950 Aveling-Barford Limited introduced their only pedestrian-controlled dumper — a 3-wheeler known as the 'Dumpling'. It was powered by a 5 hp JAP petrol engine and remained in production for eight years.

39B: The first design by the Thwaites Engineering Company Limited, a 15-cwt model announced in 1951, was also of 3-wheeled layout. Power was supplied by a 9 hp Petter engine mounted transversely beside the operator, and although there were both forward and reverse gears, no conventional gearbox was fitted.

39C: The British 'Trusty' range is less well known. This is manufactured by Tractors (London) Limited and features a JAP petrol engine. Gravity tipping is employed, with the operator seated centrally behind the dump body.

39A

39C

40A

40B

40D

40A : The Muir-Hill 3S was announced in 1956. It was a 30-cu ft model and one of the last light site dumpers built by that Company. Since then the now discontinued Winget/Muir-Hill range has catered for this market.

40B : The Thwaites 'All Drive' range is believed to have been the first site dumper to feature articulation, hydraulic steering and four-wheel drive. Introduced as a 2-tonner in 1961, the series has established the Company as one of the leading manufacturers of this type of equipment. The machine shown is a 3000 model, with a 1¼-cu yd heaped capacity, up to 12 mph forward or reverse speed and a 25-ft turning circle. An 11 hp Petter BA1 single-cylinder petrol engine supplied power to the heavy-duty four forward and four reverse constant-mesh transmission.

40C : The first four-wheel drive hydraulic tip machine to leave Barford of Belton's works was the '200', which made its début at the 1968 Public Works & Municipal Exhibition.

40D : One of the productions offered through the Hungarian National Export Company, Mogurt, is this light dumper. Types where seat and controls are reversible are frequently referred to as 'shuttle dumpers'.

41A : The French Sambron Company have been leaders in the manufacture of building and construction industry mechanical handling aids for a number of years, notably with their rough-riding fork-lift rigs. They have recently entered the UK market, one of the spearheads of this attack being their GM20 dumper with demountable 2-ton capacity skip.

41A

ON/OFF-HIGHWAY DUMPTRUCKS

The on/off-highway dumptruck concept originated in the United States during the late Twenties and early Thirties, when fleets of mainly second-hand trucks were pressed into service on the many large engineering projects then taking place. In many instances these were fitted with tipping bodies with tailgates removed, but gradually the scow-end type dump body became popular and, as the chassis wore out, so the bodies were transferred from truck to truck.

By the middle Thirties completely new dumptrucks were being assembled in a number of factories, although even here virtually standard commercial chassis were employed. In Britain the introduction of the first Foden dumptruck, in 1948, heralded the departure from the converted truck chassis to the specialist model for on/off-highway haulage.

The year 1966 saw a revision in the British Construction & Use

Regulations resulting in many types becoming illegal for road use virtually overnight. Two new companies, Heathfield Engineering Limited and Haulamatic Limited, introduced new models to conform with these revised regulations but, in July 1971, the gross weight of heavy dumptrucks on British roads was increased from 25 to 50 tons, leaving the market wide open for the bigger types to muscle in.

43B

43A

43C

42A : Typical of scow-end types used in the United States before World War II was this 1934 vintage GMC tandem-drive model. Note in particular the double-skinned cab roof and heavy front bumper.

43A : Fodens' first dumptruck was also a tandem-drive model, powered by the Company's own FED6 2-stroke diesel of 126 bhp driving through an 8-speed transmission. The all-steel scow-end body, similar to modern Foden dump bodies, had a capacity of 9/10 cu yd.

43B : Scammell Lorries Limited, of Watford, are well known for their heavy dumptruck models, all of which are now discontinued. An unusual variation was a small capacity type based on the short-wheelbase 'Highwayman' chassis. It is believed that only one such vehicle was ever built.

43C : Japanese manufacturers offer a variety of dumptrucks in this class. Hino Motors Limited, for instance, can supply the ZG 13, a half-cab type powered by a 175-bhp 6-cylinder diesel engine. The austere half-cab design is especially popular for this class of machine because of the ease with which one can replace any damaged panels.

678 HO

44A : The Thornycroft 10/13-ton dumptruck was sold in only very small numbers. It was offered in single-axle drive form or as a 4 × 4 machine, a variety of body makes being available. Power was supplied by a Thornycroft Q6 diesel engine of 170 bhp and the example shown fitted with an Edbro body.

45A : The prototype of the Haulamatic dumptruck was based on a 9 ft 4 in wheelbase Commer tipper chassis/cab, specially modified for rough site work. The revised specification included strengthened chassis, heavier duty suspension and an Allison 6-speed automatic transmission system. Early production models, still using the Commer chassis, and known as the G.P.8 or QM.8, carried Haulamatic's own half-cab design. By 1969 the series had been replaced by the 4-10, built entirely by Haulamatic Limited.

45B : The first Heathfield in this class was the DF2D 'Transdumper', a 10-ton payload machine (6 ton on the road) powered by a Perkins 6.354 diesel engine. Based on a 9-ft wheelbase single-drive chassis, common also to two larger versions of the 'DF'-Series, the DF20 was designed and built entirely by Heathfield Engineering Limited, Newton Abbot, Devon, and is capable of maintaining on-highway speeds of up to 40 mph.

45C : The Magirus-Deutz 230 D 26 AK six-wheel drive dumptruck is representative of Magirus-Deutz models offered in the United Kingdom. This 230-bhp air-cooled truck, with a 12-cu yd body capacity, is already gaining popularity with British operators.

45A

45B

45C

OFF-HIGHWAY DUMPTRUCKS

46D

46A

46B

46C

46A : The Caterpillar RD-8 crawler was especially popular during the Thirties. This one, owned by Harry T. Campbell Sons & Company, of Towsen, Maryland, is seen on a major road-building project in 1936. The two crawler dump trailers were both products of the Euclid Road Machinery Company.

46B : The 90 hp rubber-tyred Cat DW10 tractor, with a speed range from 2.4 to 18.1 mph, working in combination with the manufacturer's W10 dump trailer, was frequently employed on heavy earthmoving projects. Technical features included a 5-speed constant-mesh transmission, vacuum hydraulic and mechanical braking systems, and a ball-and-socket type drawbar coupling.

46C : The Motor Rail MR450 heavy-duty 'shuttle dumper' was a heavier version of the light site dumper, designed principally for off-highway duties. Motor Rail Limited, of Bedford, were at one time particularly well known for their rail-borne site dumpers.

46D : In 1955, the National Coal Board's Acorn Bank opencast site in Northumberland took delivery of eight trailers and four tractors built by Eagle Engineering Company Limited and Euclid (Great Britain) Limited, respectively. These were in the livery of Costain Mining Limited (Westminster Plant Company Limited), with tractors based on the popular Euclid B1TD 22-ton dump-truck design, resulting in a 53 ft 4 in overall length. Each was powered by a supercharged Rolls-Royce C6SFL diesel of 250 bhp driving through a 10-speed transmission.

The introduction of the Holt crawler had its effect upon all fields of earthmoving. One application was as a towing unit for 'double bottom' dump trailer outfits, mounted on heavy steel wheels or crawler tracks. Later, these were hauled by Caterpillar or similar combustion-engined crawlers.

Rubber-tyred equipment made its début in this field in the early Thirties when the Euclid Road Machinery Company, founded in 1931 as a division of the Euclid Crane & Hoist Company, and now trading under the name of Terex, developed a series of heavy off-highway trucks known as 'Trac-Trucs'. Initially, there was one rigid and one articulated version,

for 15 and 20 tons load capacity respectively.

Pioneers in the articulated field were undoubtedly LeTourneau and Caterpillar, Robert G. LeTourneau being responsible for the design of a unique 2-wheeled tractor dependent upon its trailer for stability.

Modern off-highway trucks are invariably 2-axle jobs with single-axle drive, notable exceptions being those with electric hub motors. These are normally four-wheel drive. For even larger capacities, certain of these rigid types have been adapted for use as tractor units, operating in conjunction with huge dump semi-trailers.

47A

47B

47C

47A : The Finnish Sisu K-36 four-wheel drive model made its first appearance in 1956 as a heavy mine dumper. In common with other off-highway types, it was equipped with an extra-wide scow-end body best suited to this kind of work.
47B : In January 1958 Fodens announced their first 56,000 lb payload dumptruck, the FR.6/45. Power was provided by a turbocharged Rolls-Royce engine developing some 300 bhp and driving through a torque converter transmission. Production figures were low—no more than one a month for a period of twenty months. It was, unfortunately, a little ahead of its time, as many of the crushers then operating in British quarries were unable to take such large capacities as 18 cu yd at any one time. The model was dropped in 1960.
47C : The 1961 International Construction Equipment Exhibition at the Crystal Palace featured this unusual machine for the first time. It was a front-wheel drive articulated dumptruck designed and manufactured by Northfield Engineering. Its payload rating was 11 tons and with power-assisted steering as standard turning circle was a mere 18 ft 9 in.

48A : Whitlock Bros, also experimenting with a front-wheel drive articulated model, introduced the strange looking DD105. It had unique hydrostatic drive to all four wheels, a 78° angle of tip and operator's cabin so placed as to provide maximum all-round visibility. Rapid advances in the use of hydraulics brought production to an end after only a few examples had been built.

48B : The Chichiang Gears Plant, in Chungking, South-West China, introduced its 25-ton 'Red Crag' dumptruck in 1966. As usual, a fairly standard rigid layout was employed.

48C : R. Smith (Horley) Limited, acquired by Matbro Limited early in 1970, had introduced their first 'Goose' dumper a few years before. Known as the 'Goose Super Four', motive power was provided by a four-wheel drive County tractor (based on the Fordson agricultural series), with suitably modified rear end designed to take the 'goose-neck' (hence the name) of the 10-ton rear-dump semi-trailer. Tipping was carried out by twin double-acting hydraulic rams driven by an hydraulic pump incorporated in the tractor transmission case.

48D : The Soviet Belaz is already well established in the off-highway class. It forms part of a range of specialist models from 27 to 110 tons capacity, announced by the Byelorussian Auto Works in 1964. Outside the Soviet Union only the 27-tonner is currently available. The prototype 65-ton model shown here was an articulated design utilizing the 27-ton chassis/cab as a prime-mover. Power was provided by a 12-cylinder 375-bhp diesel engine common to all models in the series.

48A

48C

48B

48D

OFF-HIGHWAY DUMPTRUCKS

49A : Whilst the rigid type forms the basis of current Terex dumptruck production in the UK, other designs are available. Typical is this bottom-discharge articulated outfit, known as the B-20. With a 13-cu yd struck capacity and top speed of 28 mph, the B-20 incorporates mechanical operation of the bottom-dump doors via a cable and winch system, the mechanism being operated through an air-control valve on the steering column.

49B : Practically all the leading construction equipment manufacturers can supply rigid dumptrucks in the extra heavyweight class. Representative of these is the 4 × 4 International Harvester '180 Payhauler', powered by a 2-cycle 475-bhp Detroit diesel V-12 driving through a 10-speed torque converter transmission. With a rated payload of 45 tons, struck capacity of 30 cu yd and heaped capacity of 38 cu yd, an unusual feature of this machine is the twin-tyred front axle designed to carry a gross weight almost equal to that of the rear axle.

49C : The Model 150B 'HAULPAK' truck by WABCO, powered by a GM 16V-149T 1325-bhp 16-cylinder 2-stroke diesel, is one of the largest rigid off-highway trucks in the world. Electric hub motors provide traction at all wheels and the special variable rate hydro-pneumatic 'HYDRAIR' suspension system is an interesting feature. The tail light clusters give some idea of the size of this machine, which has a heaped capacity of almost 97 cu yd.

49A

49B

49C

Grader design has remained virtually the same since 1885 when an American, Mr J. D. Adams, invented the 'leaning wheel' principle, applying it to a horse-draught grader. This 'leaning wheel' system was specially devised by Adams in order to hold the grader against skidding out of line while ditching, bank-cutting, or on other work of a similar nature. The grader was also one of the first road-building machines to be crawler-powered, one of the most successful being the Russell elevating type, built in Minneapolis by the Russell Grader Manufacturing Company, which both excavated and tipped the soil in a single pass.

With the Holt petrol crawler so popular, it was only natural that Holt should produce his own model—known as the 'Land Leveler'—and production was rationalized still further with the acquisition by the newly formed Caterpillar Tractor Co. of the successful Russell concern in 1928. Consequently, a new range of 4-wheeled towed graders was announced.

The introduction of the high-speed earthmoving sector in 1932 led to the gradual replacement of towed units by new rubber-tyred self-propelled models. The earlier types can still be found, however, and certain manufacturers still build them, principally for landscaping work or for export to under-developed countries.

51A

51B

50A : The Russell elevating grader was one of the most popular of its day. Motive power is provided here by a Caterpillar 'Sixty' crawler.

51A : Holt's 'Land Leveler' was of an entirely different design to that of the Russell. The Holt system utilized a single-axle grader with operator located on a high platform overlooking the blade. By turning a large hand wheel he could vary the depth of cut. In this instance a 5-ton Holt provided power.

51B : The Caterpillar Tractor Co.'s 8-ft grader was of more conventional layout, not unlike modern types, but still on steel wheels. The Caterpillar 'Thirty' tractor, like the larger 'Sixty', was originally of Best design and first appeared in 1921. Production ceased in 1932.

51C : The steel-wheeled D-241, a medium-duty 'terrace' grader manufactured in the Soviet Union, is still very popular, particularly for landscaping work. It is designed for use with a 54—75 hp tractor and a 3057-mm cutting edge is standard. The machine can be supplied with 9.00—20 pneumatic tyres if specified.

51C

MOTOR GRADERS

Although the self-propelled grader was a development of the trend towards high-speed earthmoving, the world's first, a 4-wheeled design built by the W. A. Riddel Corporation (now part of Huber-Warco), appeared in 1921, just over ten years before this trend became apparent. The first successful production machine was built by Caterpillar and called the 'Auto Patrol', appearing some ten years later.

By 1938 Caterpillar had introduced a heavier 6-wheeled range, follow-

ing the introduction of similar machines by LeTourneau, and it was just a matter of time before British manufacturers, such as Blaw-Knox and Aveling, were to follow suit.

A series of special attachments became available, including excavators, 'dozer blades, rippers, etc., enlarging the motor grader's scope still more, and by the 1960's four- and six-wheel drive, four-wheel steer and, later, hydrostatic transmission systems, were also applied.

52A

52B

52A : The world's first motor grader was designed and built by the American W. A. Riddel Corporation. It was a 4-wheeled solid-tyred machine with rear-mounted engine and virtually non-existent operator accommodation. It was not successful and failed to attain full production stage.
52B : In July 1925 the Wehr Company, of Milwaukee, Wisconsin,

announced a new side-cranking device for use with its power grader. This made engine cranking much easier as the Fordson power unit was mounted low between the main sideframes with the radiator and, consequently, the cranking point located inaccessibly immediately to the rear of the operator's feet. The new system was ordinarily operated from the right-hand side.

MOTOR GRADERS

53A : The first motor grader to reach full production stage was the Caterpillar 'Auto Patrol' of 1931. Designed principally for highway maintenance, features included a rear engine, transmission direct to rear wheels eliminating all prop shafts, a cantilevered frame, and a special scarifier and ripper system ahead of the blade.

53B : A heavy-duty 6-wheeled version was introduced by Caterpillar in 1938. This was aimed at the rapidly expanding construction and earth-moving industries, new features including four-wheel drive and a fully-enclosed cab to eliminate some of the dust problems. Blade control was still through a mechanical system although much of this was also dust-sealed.

53C : The British-built Blaw-Knox BK-12 was introduced in November 1949. Equipped with a 12-ft blade and powered by a 7.7-litre 94 hp AEC oil engine, it was designed to tackle the toughest of jobs as indicated by its weight of $9\frac{1}{2}$ tons. It was possible, by this time, to swing the blade still further away from a horizontal plane so that bank levelling and sloping could also be undertaken.

53A

53B

53C

54A : The Blaw-Knox 'Super 12', a development of the BK-12, was offered with a special elevator system for side-dumping of graded material. This was particularly useful when constructing a raised road surface using material from the roadside. Movements were mechanically controlled with elevator luffing by hand winch.

54B : The Adams '50' motor grader, built by the Adams Manufacturing Company, Indianapolis, Indiana, was the forerunner of the modern WABCO grader, built by WABCO/CED. Adams was purchased outright by the old LeTourneau-Westinghouse Company in December 1954.

54C : The Galion motor grader is available in a number of countries with a variety of power units, depending upon availability in the country of operation. This version is powered by Scania and features hydraulic blade control, both for grader and 'dozer blades. An unusual feature is the very long wheelbase which permits the grader blade (also known as the moldboard) to be completely reversed for operation in either direction.

54D : In 1937 the Austin-Western Company of Aurora, Illinois, introduced their first all-wheel drive and all-wheel steer motor grader. To meet the needs of world markets, Aveling-Barford Limited, of Grantham, Lincolnshire, in association with Austin-Western, commenced production of the 99-H machine in 1950. The current 99-H shown here, a development of the original model, features a 150 hp Leyland or, alternatively, 128 hp Detroit diesel engine coupled with full hydraulic control. The cabin is an optional extra.

54B

54C

54D

54A

55A : The Aveling-Barford maintenance grader, introduced in 1954, was an adaptation of the Fordson diesel industrial tractor. It was in production for nine years, featuring full hydraulic control of all blade movements.

55B : Currently the most popular machine in the Wakefield range, produced by British Jeffrey-Diamond Limited, is the 'Wakefield 130'. This was announced in 1965 as an improvement on the earlier '120'. The '130' was first shown at the International Construction Equipment Exhibition, Crystal Palace, in that year and is regarded by many as one of the most advanced machines of its day. It is powered by a 130 hp Leyland diesel, driving through a 6-speed transmission to the chain-drive tandem rear bogie.

55C : The Soviet Belaz D598-E is built along virtually identical lines to models produced by western manufacturers. The operator's cabin is fully weatherproofed and dust-sealed, and the machine is powered by a 75 hp 4-cylinder diesel driving through a mechanical transmission providing 6 forward and 2 reverse speeds. The blade is fully reversible by means of a hydraulic motor and can be fitted with a blade extension if necessary.

55B

55A

55C

TOWED SCRAPERS

56A : The early towed scraper unit was developed during the 1870's, principally for use on the big railroad projects of the American Midwest. Instead of carrying the material to the tip area, these heavy steel-wheeled types dragged the spoil to the dumping ground.

57A : The Maney 4-wheeled model was a development of the early 1900's. It cut the cost of dirt-moving by nearly 50% and was available in capacities of up to 1½ cu yd.

57B : Robert G. LeTourneau entered the earthmoving industry as a land-levelling contractor. As he was anxious to obtain greater efficiency, he set about designing his own equipment. His first production model was called a 'Gondola', welded throughout and much lighter in weight than comparable rigs of the day. Its heaped capacity was approximately 6 cu yd and it is interesting to note that, even at this early date, electric motors were used to tilt the bowl for loading and unloading.

57C : In an attempt to solve flotation problems, LeTourneau developed a track-type tread scraper in 1926. This was also electrically operated with the front of the scraper moved up and down on rachets to control digging actions.

By the late 18th Century man was beginning to realize that for the larger earthmoving contracts the shovel and barrow were far from practical. Thus, about 1805 the first scraper was evolved, a small scoop, pulled by horses and controlled by a worker, the most popular models being the Dragpan and the Fresno. By the 1870's these had evolved into the first wheeled scrapers, heavy cumbrous appliances capable of shifting 100 to 250 cu yd per day, depending upon material and length of haul, although even the Fresno could still be seen with a crawler motive-unit as late as 1930.

Earthmoving contracts were now increasing in size and complexity with contractors finding even the existing aids most uneconomical and inefficient. With the advent of the crawler tractor many new possibilities were opened up. Soon after World War I, R. G. LeTourneau, for instance, invented his first scraper, for use with a Holt '75' crawler machine.

57B

57C

57A

58A

58B

58A : In 1932 a Californian contractor using steel-wheeled LeTourneau scrapers encountered some difficulty while working the machine in sand. Mr LeTourneau persuaded him to accept a rubber-tyred conversion which was an immediate success. High speed earthmoving had arrived.

58B : Early the following year LeTourneau introduced the 'Carryall' type scraper. Instead of dragging its load to the tip area, this new design actually carried the material and, later that year, an improved version (known as the 'B'-Type) was developed. All modern designs are now based on the 'B'-Type layout, which featured an apron to hold the dirt in while travelling, a 'tailgate' for ejecting the load, and operator-controlled depth of spread. The 'Carryall' was the first earthmover that could perform all three of the basic steps in earthmoving – loading, hauling and dumping.

58C : The Caterpillar 'Fifty' diesel crawler was frequently assigned to scraper duties. LeTourneau's famous power control unit was invaluable for this type of work and was, for many years, a standard fitting on many crawlers. This night operation was on the Treasure Island approach to the San Francisco Bay Bridge, 1938.

58D : The Vickers Onions towed scraper was especially popular during the Fifties and Sixties. This 12–16 model featured full cable control and was coupled to a Vickers 'Vigor' unit. The total power output was 360 hp, driving through a power transmission or torque converter.

58C

Another development was to couple two or more scrapers together. With the power available from these crawlers an operator could not, of course, load all bowls at once but had instead to load each in turn. This somewhat defeated the object, but still it was a step in the right direction.

Various designs were perpetrated during the 1920's, notably in the United States, where earthmoving requirements were more advanced than in many other countries. 1928 saw the introduction of the first cable-controlled scraper, requiring only one operator, and in 1932 the first rubber-tyred scraper, again a LeTourneau development, entered service. The following year LeTourneau developed the first scraper designed to carry its load to the tip as opposed to earlier types that dragged the load, and consequently, lost a good deal en route. This system remains the basis of modern towed scrapers, the design of which has certainly revolutionized the earthmoving industry.

58D

59A : As the 'teething problems' were ironed out, so less complicated, and larger, scrapers were evolved. The Blaw-Knox BK-80 was one of these, based on similar lines to American models. This unit is spreading its load via the forward door by winching the tailgate towards the front of the bowl.

59B : At times, operating conditions can be extremely arduous if not diabolical. Typical of many sites is this one, where the mud adheres to everything. The scraper is a 21 cu yd (28 cu yd heaped) Cat 463 model hauled by a Caterpillar diesel tractor.

59C : Here, another 463 scraper loads on the Gulf Oil Terminal Contract at Milford Haven, South Wales. Although the 463 is the largest towed scraper currently offered by the Caterpillar Tractor Co., the D8H crawler seen here has power to spare for this type of work.

59B

59A

59C

MOTOR SCRAPERS

LeTourneau again came to the fore in 1938 when his first motorized scraper appeared. He designed this in 1937, basing his idea on a single-axle prime mover relying upon the scraper bowl for its even balance and stability. He utilized a much modified Caterpillar crawler tractor as a basis for early development models, incorporating the crawler tractor's original steering mechanism, by de-clutching and braking one wheel through an interconnected gear. Tyres were produced to order by the Firestone Tire & Rubber Company.

The earthmoving industry was amazed at his ingenuity and the new model, known as the 'Tournapull', was soon in quantity production. Demand, in fact, was so great that R. G. LeTourneau Inc. had become the leading manufacturer of earthmoving equipment by the start of World War II, supplying an estimated 70% of all basic earthmoving equipment used by the world's armed forces during the years of conflict. Other manufacturers were quick to realize the potential of these machines and within no time at all there was a host of models available.

During the latter half of 1945 R. G. LeTourneau investigated the possi-

60A

60C

60B

60D

60A : The Model 'C Tournapull' was the world's first fully self-contained motor scraper. This appeared in 1938, utilizing the same scraper bowl as for LeTourneau's towed models. The coupling of tractor and trailer permitted sideways rocking of the tractor, whilst maintaining rigidity in a fore-and-aft direction. Traction was excellent, with much of the weight applied to the driven tractor wheels. Not only did this model speed up the earthmoving cycle, but it also made the whole more economical. The pusher unit, a Caterpillar crawler tractor with rooter attachment, continued with the task of rooting and fragmentizing hard-packed material while the scraper off-loaded.

60B : The 'Super C Tournapull' was announced in 1940 as a 12-cu yd model powered by a 150 hp engine. This machine became the standard for

high-speed earthmoving for many years, giving excellent service throughout the war.

60C : The Blaw-Knox 'Goliath' was not dissimilar to LeTourneau's design, utilizing a 2-wheeled tractor obtaining its stability from the trailer via an upright kingpin. It was of 8-cu yd capacity and the first British model to employ a single-axle tractor, with power-assisted steering effected by a pair of hydraulic rams.

60D : The Euclid Road Machinery Company did not, at first, favour LeTourneau's single-axle tractor design. They preferred the more conventional 2-axle unit with parts interchangeable with their other off-highway types.

bility of using an electric control system for scraper movements. Following numerous experiments and pre-production tests, the new system was made available in 1947 and, in the same year, the 'Super C Tournapull' was dropped from the Company's product line to be replaced by the latest electric control model.

From 7 cu yd before the war, capacities had increased to 15 and more after, with engine power also rising. By the late Forties both Euclid and Caterpillar were offering competitive models and Blaw-Knox announced their first type, the 'Goliath', in 1948.

The first production twin-engined models by the Euclid Road Machinery Company, made their début in 1949 and the introduction of full hydraulic control enabled even larger types, some up to 40 cu yd capacity, to become practicable. It should be remembered, however, that as early as 1942 R. G. LeTourneau had built a twin-engined model, with engines mounted side by side at the front of the machine. This gave a total output of 400 hp!

61A

61B

61C

61A : For extra capacity the LeTourneau-Westinghouse Corporation (as it was known from 1953) introduced the C500 'Pushpak', consisting of two scraper bowls and twin engines one at the front and one at the rear. The front bowl would be loaded first and the trailing unit second.

61B : The Terex TS-14 and TS-24 scrapers are twin-engined types with capacities of 14 and 24 cu yd respectively. Euclid, the forerunner of Terex, developed the twin-engine system during the very early 1950's. Now the majority of heavy plant manufacturers offer such machines as standard. Terex can also supply a 3-engined version of the TS-14 (known as the 'Tandem' TS-14), the largest scraper in their range. This features a second TS-14 bowl, with engine, coupled to the rear of the first. These TS-24's were used on site clearance for the M5 Mendip Motorway in 1970.

61C : The self-loading elevating scraper was developed in the United States by the Hancock Manufacturing Company during the early Fifties, the first production model, a 5-cu yd design, appearing in 1952. The idea was to increase load cycle efficiency. Since then Hancock have, in many cases, produced the complete machine (including motive unit), although many of their scraper designs have been coupled to other makes of tractor. In 1966 Hancock merged with the Clark Equipment Company, which had produced its first motor scraper in 1957, to develop an even broader range of self-loaders. Representative of these is the Michigan Model 110 Series III which can cope with a 14-cu yd heaped load. A 238 hp engine provides the power.

62A : Even twin-engined types require assistance on occasions. This 950 hp Caterpillar 657 is 'banked' by two Cat D9G machines, the first with single 'dozer ram and the second with the more conventional twin-ram layout. The 657, a 32-cu yd machine capable of carrying a 44-cu yd heaped load, has a semi-automatic transmission providing 8 forward and 1 reverse speed. All scraper controls feature a double-acting hydraulic system. A new idea, recently introduced into the Caterpillar Tractor Co.'s product range, is the push/pull system, where two Cat twin-engined outfits are specially constructed for pushing or pulling each other while loading, hauling and dumping.

63A : The WABCO 229F is a direct descendant of the original 'C' and 'Super C Tournapull' machines. The 229F, however, features a GM 8V-71N engine of 318 hp driving through an Allison power-shift transmission. Capacity is 15 cu yd struck or 21 cu yd heaped.

63B : The 'Goose' elevating scraper incorporates a system similar to Hancock's design. The motive unit is a County 4 × 4 tractor, and the scraper equipment, originally developed by R. Smith (Horley) Limited, is now manufactured by Matbro Limited.

63C : WABCO's latest self-loader, announced early in 1971, is the 101F, a 9-cu yd machine powered by a de-rated Cummins V-504 V-8 diesel, developing 178 hp. This is a comparatively small machine, a special feature of this design being the 'command post' operator's cockpit above the engine.

63B

63A

63C

INDEX

ACKNOWLEDGEMENTS

Credit is hereby given to all manufacturers, operators and other bodies who have assisted in the compilation of this book. In particular, we should like to thank the American Hoist & Derrick Company, Blaw-Knox Limited, the Caterpillar Tractor Company and Caterpillar Overseas S.A., Hy-Mac Limited, the International Harvester Company, Anderson Mavor Limited, Poclain, General Motors Scotland Limited, the Westinghouse Air Brake Company, the Commercial Vehicle & Road Transport Club, the B. H. Vanderveen Collection and Mr Larry Auten.